MONARCH BUTTERFLIES

A MIGRATION STORY

by Lisa Amstutz ▪ illustrated by Howard Gray

AMICUS ILLUSTRATED is published by
Amicus Learning, an imprint of Amcius
P.O. Box 227, Mankato, MN 56002
www.amicuspublishing.us

Editor: Alissa Thielges
Series Designer: Kim Pfeffer
Book Designer: Emily Dietz

Library of Congress Cataloging-in-Publication Data
Names: Amstutz, Lisa J., author. | Gray, Howard, illustrator.
Title: Monarch butterflies : a migration story / by Lisa Amstutz ; illustrated by Howard Gray.
Description: Mankato, MN : Amicus Illustrated, [2026] | Series: Incredible migrations | Includes bibliographical references. | Audience: Ages 5–10 | Audience: Grades 2–3 | Summary: "Follow the incredible journey of monarch butterflies' migration from the eastern United States to Mexico in this narrative nonfiction picture book that will delight animal and nature lovers and support life science education. Includes a migration map, tips to protect animals and habitats, a glossary, and further resources"—Provided by publisher.
Identifiers: LCCN 2024043391 (print) | LCCN 2024043392 (ebook) | ISBN 9781645492849 (library binding) | ISBN 9781681528083 (paperback) | ISBN 9781645493723 (ebook)
Subjects: LCSH: Monarch butterfly—Migration—Juvenile literature. | Monarch butterfly—Behavior—Juvenile literature.
Classification: LCC QL561.D3 A497 2026 (print) | LCC QL561.D3 (ebook) | DDC 595.78/9—dc23/eng/20250103
LC record available at https://lccn.loc.gov/2024043391
LC ebook record available at https://lccn.loc.gov/2024043392

About the Author

Lisa Amstutz is the author of more than 150 children's books. A former outdoor educator, she holds degrees in biology and environmental science. Lisa enjoys learning fun facts about science and sharing them with kids. She lives on a small farm with her family.

About the Illustrator

Howard Gray has illustrated a selection of fiction and non-fiction children's books. He has always considered himself an artist, but with a PhD in dolphin genetics, he has a background in zoology. He is now pursuing his dream career in children's illustration from the picturesque city of Durham, UK. Find out more at www.howardgrayillustrations.com.

Stretch! A monarch butterfly spreads her wings. The August sun is warm. But fall is on its way. It is time to fly south! She must migrate to a warmer place. She will travel up to 3,000 miles (4,800 kilometers) from her home in eastern North America.

The wind lifts the butterfly high. She soars on air currents. She can travel 50 miles (80 km) per day. But she is not alone. Many other monarchs follow the same path.

The butterfly flies all day. At night, she gathers on a tree to rest with the others.

The next day, the butterfly sets off. But rain begins to fall. *Plip, plop!* The water makes her body heavy. It can damage her wings. She takes shelter under a leaf.

When the sun comes out, she dries her wings. She sips nectar from a flower. Then she lifts off again. The butterfly has never migrated before. But she knows how to find the way. The earth has a magnetic field. She uses it like a compass to know which way to fly. She uses the sun, too.

It is a dangerous trip. Milkweed sap makes monarchs taste bitter. But some birds will eat them anyway.

People spray pesticides to get rid of weeds and bugs. They mow ditches and pave fields. Flowers are getting harder to find.

The butterfly soars over rivers and lakes. She crosses mountains and deserts. In November, she reaches central Mexico. This will be her winter home. Monarchs in western North America fly to California.

The butterfly roosts on a fir tree. Millions of others land nearby. They huddle to keep warm. Thick branches shelter them from the wind.

In March, the butterfly finds a mate. Then she starts the journey back home. She lays hundreds of eggs along the way. But she will not make it home. She is only eight months old, but monarchs do not live long. Her job is done. The next three generations will finish the trip.

In May, a small caterpillar hatches. It feeds for about two weeks. Then it makes a pupa. In 8 to 15 days, it turns into a new butterfly.

It too heads north, laying eggs on milkweed plants. Its children and grandchildren will do the same. These generations live about four weeks each.

At last, a butterfly reaches its great-grandmother's home. She lays tiny eggs. In August, they hatch. Soon, they will spread their wings and fly . . .

all the way to Mexico!

MONARCH BUTTERFLY MIGRATION

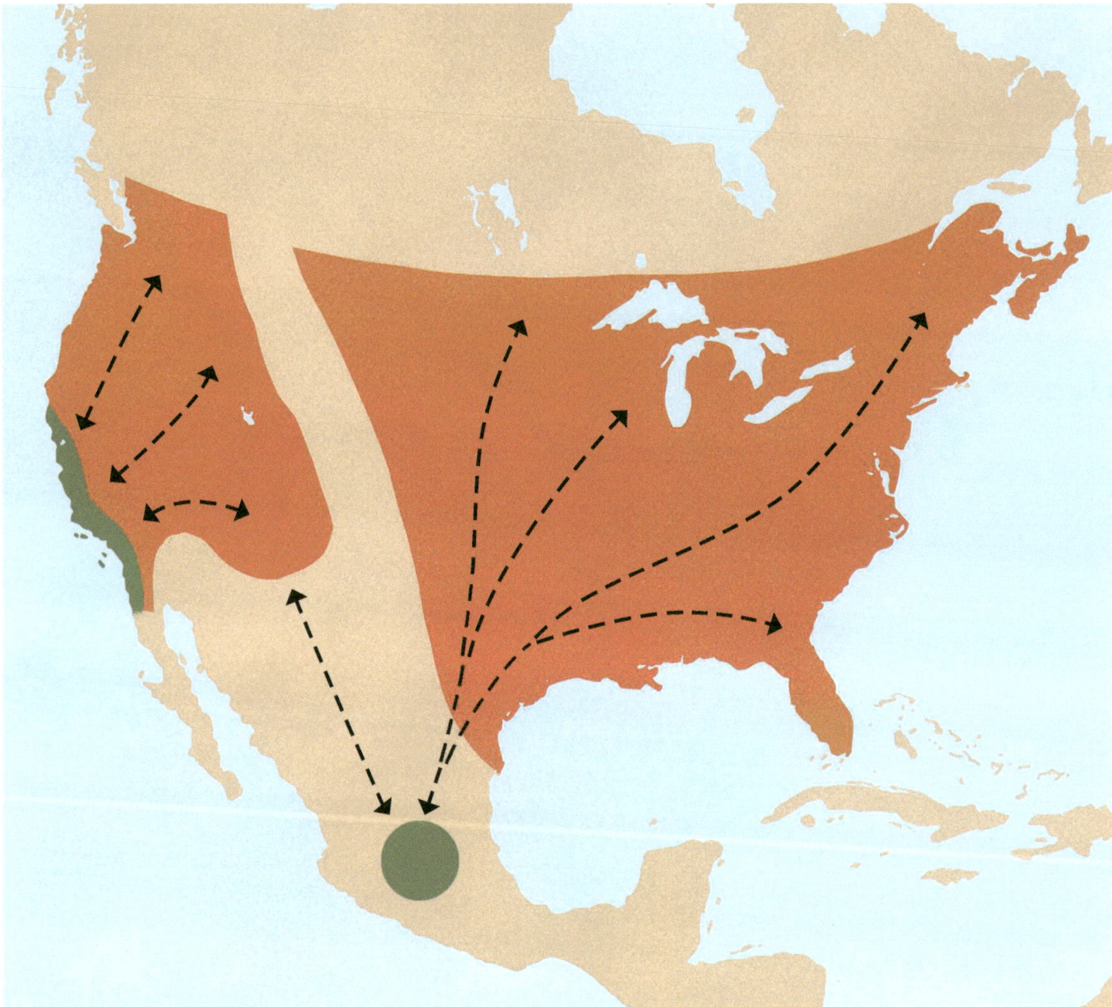

Spring/Summer

Winter

Migration routes

HOW CAN I HELP ?

Monarch butterflies face many threats. Here's how you can help!

- Plant native wildflowers for adult butterflies to feed on.
- Do not mow or spray your ditches during monarch season.
- Put out a shallow dish of water for butterflies to drink. Add stones for them to sit on.
- Plant milkweed in your garden or flowerbeds. Monarch caterpillars feed only on this plant.
- Spot a monarch? Report its location online! Sites like Journey North help scientists track the butterflies' movements. Go to https://journeynorth.org/monarchs.

Glossary

current Air moving in a certain direction.

generation All the members of a group born around the same time.

mate The female or male partner of a pair of animals.

migrate To move from one place to another to breed or find food.

nectar A sweet liquid that some insects collect from flowers and eat as food.

pesticide Poisonous chemicals used to kill insects.

pupa A hard casing with an animal inside; the animal is changing from larval stage to the final animal stage.

roost To settle in a group to rest.

Read More

Banks, Rosie. ***Why Do Animals Migrate?*** New York: Gareth Stevens Publishing, 2024.

Donnelly, Rebecca. ***On the Move with Monarch Butterflies.*** Minneapolis: Jump! Inc., 2023.

Pryor, Katherine. ***Home Is Calling: The Journey of the Monarch Butterfly.*** New York: WorthyKids/Hachette Book Group, 2023.

Websites

Decoding the Great Monarch Butterfly Migration
https://thekidshouldseethis.com/post/decoding-the-great-monarch-butterfly-migration

National Geographic Kids | Monarch Butterfly
https://kids.nationalgeographic.com/animals/invertebrates/facts/monarch-butterfly